BR BLUE

BR BLUE

SCENES FROM THE BRITISH RAIL CORPORATE IMAGE ERA

MARTYN HILBERT

FONTHILL

Fonthill Media Language Policy

Fonthill Media publishes in the international English language market. One language edition is published worldwide. As there are minor differences in spelling and presentation, especially with regard to American English and British English, a policy is necessary to define which form of English to use. The Fonthill Policy is to use the form of English native to the author. Martyn Hilbert was born and educated in England; therefore, British English has been adopted in this publication.

Fonthill Media Limited
Fonthill Media LLC
www.fonthillmedia.com
office@fonthillmedia.com

First published in the United Kingdom and the United States of America 2022

British Library Cataloguing in Publication Data:
A catalogue record for this book is available from the British Library

ISBN 978-1-78155-864-5

Typeset in 10pt on 13pt Sabon
Printed and bound in England

Contents

Introduction 7

1 London Midland Region 15

2 Eastern Region 35

3 Southern Region 51

4 Western Region 63

5 Scottish Region 77

6 BR Blue Miscellany 85

7 Resurgence and Epilogue 95

Bibliography 96

Introduction

Following the Transport Act of 1947, the railway system in Britain was nationalised in January 1948. The newly created 'British Railways' was divided into six regional areas, some of which broadly covered the operating areas of the previous 'Big Four' railway companies. The Western Region, Southern Region, London Midland Region, Eastern Region, North-Eastern Region, and Scottish Region all had their own identities with their coloured signage and were controlled by regional management, overseen by the British Transport Commission (BTC) and the British Railways Board (BRB). While conforming to policies laid down by headquarters, many of the regions tended to 'do their own thing', and following the 1955 Modernisation Plan, in the years that followed, a plethora of new diesel locomotives appeared on the system from not only BR's own workshops, but also from a variety of outside manufacturers. In many cases, bulk orders were given for a particular type without any prototypes being built to assess their reliability or suitability.

Between 1955 and 1963, some 4,600 individual diesel multiple unit (DMU) cars were also introduced, which changed the face of BR operations across the country. By the early 1960s and with ongoing costs/losses running out of control, the government of the day commissioned the notorious Beeching Report of 1963, the results of which set in motion a number of initiatives, the legacies of which can still be seen on the remaining network today.

The BR design panel was created to give British Railways a new image in the early 1960s during a period when Britain was undergoing radical social and economic changes. The *BR Corporate Image Manual* was initially issued in July 1965, with additional updates and revised sections added in November 1966 and April 1970. The manual laid down how the 'new' British Rail would look. A new logo was devised, by Gerry Barney, who was part of the design research unit—the double arrow symbol (still in use for station entrance signage today)—and a new signage scheme was initiated with specially designed typeface: 'rail alphabet' in upper and lower case and in either black or white depending on the use. The new style 'British Rail' lettering was devised by Margaret Calvert and Jock Kinnear, both of whom had devised the new signage for the UK road and motorway networks.

The BR logo and rail alphabet still in use at Goxhill in North Lincolnshire in 2021.

Four new colours were selected to give the British Rail system a house style and a clean and modern image—rail blue, rail grey, flame red, and marine blue. The *Corporate Image Manual* detailed how locomotives, multiple units, rolling stock, road vehicles, Freightliner, Motorail, Railair, Rail Drive, InterCity, rail freight, rail express parcels, stations, hotels, buffets, ships, and hovercraft should be presented. The detail was immense and among the items covered were literature, menus, posters, linen, uniforms, catering packaging, tableware, carpets, research, British Transport Films (BTF) credits, timetables, leaflets, signage, clock face style, and even litter bins.

The first complete trains to wear the new 'rail blue' livery were the forty-three ex-London Underground tube cars that were refurbished for use on the 8.5-mile Ryde to Shanklin line on the Isle of Wight in 1966; the trains were formed into three- and four-car (3-VEC and 4-TIS) sets when the newly electrified line opened for business in March 1967.

Elimination of steam traction across the country had been ongoing since the 1950s, but steam finally bowed out on the network in the north-west in August 1968. As the Swinging Sixties became the 1970s, there was an ongoing programme of refreshing the system, including changing station signage. Previously, in 1968, the National Traction Plan had looked to the future of the BR locomotive fleet across the network and their role in a rapidly changing system. This would see the reduction of locomotive types, standardising on diesel-electric transmission, and in particular the elimination of the many types of diesel shunters on the books, as traditional freight flows and goods yards/facilities went into decline as a result of improving lorry/truck design and the expansion of the motorway and road networks.

A major change occurred in the early 1970s with the renumbering of the locomotive fleet so that the numbers would be compatible with the TOPS (total operations processing system) via computer. The diesel and electric locomotive fleet numbering system had previously been modified in 1968, by the gradual deletion of the 'D' and 'E' prefixes to the running numbers.

By the mid-1970s, the previous 'old' liveries were getting thin on the ground, and as the locomotive fleet was in a continual state of flux, green locomotives would pass

Above: 4-VEC 043 at Ryde Esplanade with the Ryde Pier shuttle service on 25 July 1978.

Right: The front cover of the 1972 edition of Ian Allan's *British Railways Locomotives and Other Motive Power Combined Volume*. It was the first of these books to have a linen cover, and the last to feature the BR locomotive fleet with Dxxxx and Exxxx numbers prior to the introduction of TOPS renumbering the following year.

Still in BR green livery (with a blue patch where the TOPS number had applied over its old number 3380), 08310 was at Doncaster Works on 17 June 1978.

through the main workshops and emerge newly overhauled in rail blue livery. Even the three narrow-gauge steam locos and the coaches on the BR 1-foot 11.5-inch (600-mm) gauge Vale of Rheidol line in Wales succumbed to rail blue.

In 1975, the last few BR Mk I coaches that were still running in the obsolete BR maroon livery were formed into a ten-coach 'Derby Special Traffic Set' and with that exception, most of the objectives of the *Corporate Image Manual* were by and large completed around 1978–79. The last BR green-liveried diesel locomotive was Class 40 40106 (D306), and to everyone's surprise when it was overhauled at Crewe Works in 1978, it was outshopped in retro BR green with yellow ends.

In a changing world, and with different attitudes at headquarters and at local levels, slowly, some deviations from the 'norm' started to appear. For the Silver Jubilee celebrations of HM Queen Elizabeth II in 1977, Stratford depot in East London adorned two of their Class 47s (47163 and 47164) with large bodyside Union Jack flags that were complemented by painting the loco roof silver. The silver roof would eventually be applied to many Class 47s in the Stratford fold, but across the system, local custom adornments started to appear on various locomotives across the fleet—such as white stripes, white cab surrounds, and red buffer beams—giving some individuality to the system and creating some local pride. When newly applied, rail blue was quite glossy,

Class 40 40106 in its retro BR green livery that was applied at Crewe Works following overhaul in 1978. The loco was on display at the open day at Knottingley depot on 5 May 1980.

but the finish soon weathered and the appearance eventually became quite flat. It was particularly noticeable on the multiple unit fleet; from the mid-1970s, much of the fleet was painted in blue/grey livery following repairs or overhaul.

A startling new livery variation was applied to new build Class 56 56036 in 1977, with a rail blue base colour, silver roof, yellow wrap-round cabs, and a large BR logo and numbers. This went on to be known as 'large logo' livery, which was also applied to some overhauled Class 47s and Class 50s, transforming their appearance.

The introduction of the InterCity 125 HST fleet on the Western Region in 1976 gave another new livery, with the power cars having a yellow/blue/grey livery that blended with the blue/grey Mk 3 trailer coaches. With the formation of passenger transport executives (PTE) in the mid-1970s that partly funded and oversaw the running of services in the large cities, PTE logos started to be applied to specific area multiple units, initially in BR corporate image style and later followed by some regional PTE liveries and logos; Greater Manchester and Strathcylde (Glasgow) were two of the most distinctive.

As the 1970s became the 1980s, many other custom paint finishes appeared, along with other new liveries (such as Intercity, Merseyrail, Network South East, Railfreight, Scotrail, and sectorisation), as the BR system was readied for eventual privatisation, but

In large logo blue livery, Class 56 56084 was at Preston Docks awaiting departure to Lindsey oil refinery on 25 July 1984.

those liveries are a story for another day. During the late 1970s and early 1980s, many of the refurbished first-generation DMU fleet were finished in a short-lived white with blue stripe livery. This looked very smart when newly applied but was not practical for the services and in the environments that these trains operated.

The rail blue era was still evident into the 1990s with numerous locomotives and rolling stock still wearing the then rapidly becoming obsolete livery. I have tried to give a balanced coverage of the network, but in ninety-six pages, inevitably some selective compression of the regions has occurred. I have also tried to give a good selection of the broad spectrum of locomotives, multiple units, and rolling stock then in service and some of the infrastructure, which in many cases is now part of history. The BR rail blue period is the only time in the forty-eight years of the nationalised network that an identical 'corporate style' was in evidence nationwide.

In the preparation of this book, one thing that has been very apparent is that most of the hardware that is illustrated in the following pages no longer exists, having been scrapped. Some locations are still recognisable, while others have vanished altogether; while time has been not too kind to other places, in just a few, the passage of the years has been a little more sympathetic.

Unlike today, when a Class 66 diesel can be working in Kent one day and then the next day is seen in Cumbria, many locomotives generally kept to their own region and to see certain classes, one had to travel to far-flung corners of the BR network to see some locos in their natural habitat. It was all good entertainment and a great way to see the country and to experience the operations of the nationalised system and the places it served.

The rail blue era is now looked back upon as a golden age, when there was a huge variety of loco types and rolling stock operating a myriad of services and workings. The period is now one that is recreated by the preservation movement as we look back at rail blue with the same fondness that the BR green era was looked upon when BR blue was the 'norm'.

Virtually all of the following images are scanned from Kodachrome 25 and 64 transparencies using a variety of 35-mm SLR cameras, including a Zenit B and a Praktica TL, with the vast majority being done with a superb Olympus OM1 that had been purchased new from Dixons in 1977. The images cover the period from the mid-1970s into the early 1990s.

My thanks to Alan Sutton and the team at Fonthill Media, who have enabled another selection of my work to appear in print, and thanks to my wife, Gillian (she too witnessed the BR blue era). I hope that you the reader enjoy the selection of images that follow, taking a look back at a vanished world when the nationalised railway system in Britain was in the rail blue era.

Martyn Hilbert
Lostock Hall, Preston, Lancashire
June 2022

1

London Midland Region

High in the Pennines on the rooftop of England at 1,150 feet above sea level and 5 miles from the small village in purports to scrve, Class 45 45049 *The Staffordshire Regiment (The Prince of Wales's Own)* was passing though the then-closed station at Dent with a lengthy rake of empty HAA coal hoppers bound for Carlisle New Yard on 23 August 1977. No. 45049 had been new from Crewe Works, numbered D71 in November 1960. It was withdrawn from service in October 1987 and scrapped by MC Metals in Glasgow in April 1989. Having closed in 1970, Dent station was reopened in 1986.

With the up distant signal showing 'clear', Class 45 45041 *Royal Tank Regiment* was sweeping downgrade towards Dent station with the 11.50 Glasgow Central to Nottingham service formed entirely of BR Mk I coaches on 23 August 1977. The coach behind the locomotive—a BR Mk I Brake First Corridor (BFK)—looks as though it has been added to the train to increase capacity on this busy service.

Derby-built Class 115 Suburban DMU sets were stabled alongside what was once the Great Central Main Line, just to the north of Aylesbury station on 8 January 1984. The nearest vehicles were both Driving Motor Brake Second Opens with M51888 on the left, alongside M51670. There had been forty-one four-car sets introduced in 1960, mainly for use on services from London Marylebone over the former GC and GC/GWR joint lines. A small batch of 115s based at Allerton depot on Merseyside also operated services along the former Cheshire Lines (CLC) route between Liverpool and Manchester. In the twilight years of the Great Central Main Line from Marylebone to the north, 115s also worked through to Nottingham Victoria. The last of the Marylebone sets were withdrawn in 1992 following the introduction of new Class 165 Turbo DMUs.

Class 40 40099 was stood at the north end of platform 3 at Preston station during an engineering possession on Sunday, 25 September 1978. The loco had been built at Vulcan Foundry, Newton-Le-Willows, as D299 and had been new in October 1960. The 133-ton loco lasted in service just shy of twenty-four years; it was withdrawn in October 1984 and subsequently scrapped at Doncaster Works in March 1985.

Against a mountain of slate waste, a ten-car DMU formation was stabled in the former LNWR goods yard at Blaenau Ffestiniog on 25 May 1978. The lengthy DMU had arrived on a rambler's excursion from East Lancashire and had visited the North Wales seaside resort of Llandudno before traversing the 30.8-mile Conwy Valley route. The formation consisted of a BRCW three-car Class 104, a Derby two-car Class 108, a Cravens two-car Class 105 Power Twin, and another BRCW three-car Class 104 with car M50468 at the rear. All of the DMU sets were allocated to Newton Heath depot in Manchester.

Class 25/3 25324 was Sunday resting in the south-facing bay platform at Rugby station on a damp 17 May 1981. Rugby still retained its commodious former LNWR overall roof, which was eventually demolished in 2000. The Class 25 was one of the last batch built, completed at Derby Works as D7674 in March 1967. The loco should have been built by the famous Beyer-Peacock Company in Manchester, but as they went into liquidation in 1966, production of the last eighteen locomotives (D7660–D7677) was switched to Derby.

With a rather empty-looking M6 in the foreground, Class 87 87010 *King Arthur* was threading through the Lune Gorge with the 7.45 London Euston to Glasgow Central service on 12 July 1982. There were thirty-six 5,000-hp Class 87s (87001–87035 and 87101) built at Crewe Works between 1973 and 1975. They were specifically designed for working Anglo-Scottish traffic on what was then the newly electrified section of the WCML between Weaver Junction and Glasgow Central. The locos were originally designated 'AL7' and were allocated numbers in the 'E32xx' series; although the numbers were used on the production line at Crewe, all the 87s entered service with 87xxx numbers. No. 87010 was withdrawn from service in 2005 and was exported to Bulgaria in 2008.

In a time-honoured scene that was played out daily at the former Midland Railway's magnificent London terminus and is now consigned to history, Class 45 45134 with its Mk 2 air-conditioned coaches was stood in the evening sunlight at St Pancras with the 17.50 departure to Nottingham on 12 April 1980. This side of the station has since been modified/ extended for use by international Eurostar services.

With the new extension to the Fox's biscuit factory under construction, Class 31 31249 was heading for the Fylde Coast as it was passing Kirkham and Wesham with the 13.50 Manchester Victoria to Blackpool North service on 24 September 1988. A 21-ton mineral wagon, three 16-ton mineral wagons, and a long wheelbase 10-ton tube wagon all in departmental/engineers use were parked in what had been part of the goods yard.

Having arrived at Crewe with the 09.37 Llandudno to Birmingham New Street service, Class 47/4 47448 was awaiting to be replaced by an AC electric locomotive for the southbound run to the West Midlands on Saturday, 9 June 1979. The loco had been built at nearby Crewe Works as D1565 in March 1964. It was withdrawn, having completed just over twenty-seven years in service, in May 1991. No. 47448 was converted into 112 tons of scrap metal by Booth's of Rotherham in 1996.

Class 20s 20169 and 20092 were stabled in Bescot yard with a rake of HAA coal hoppers as one of the exhibits at the open day held on 30 August 1992. The pair of locos and wagons had been specially posed for the railway artist Philip Hawkins to produce a painting depicting a once familiar scene in the yard. Bescot Yard at Walsall is famously visible alongside the busy M6/M5 Motorway at junction 8. The yard once had a hump shunting facility, and in the 1960s, the yard could handle 4,000 wagons daily. As traditional wagonload freight has declined, so has activity Bescot. There is now a Network Rail 'Virtual Quarry' storing track ballast, and much of the extensive yard is now devoted to engineers' trains/wagons.

Class 37 37215 was heading eastbound passing Lostock Hall Junction with the Saturday-only 11.45 Blackpool North to Sheffield service on 1 June 1985. This summer-only out-and-back 'holiday' service ran for many years and in the 1970s and early 1980s brought the rare sight of a Class 37 on a passenger working to the Preston area. The area behind the loco and coaches was once the site of a railway wagon repair workshop and sidings.

Former Western Region Class 123 InterCity Driving Car E52094 was stood in platform 1 at Manchester Piccadilly after arrival with a service from Sheffield on 18 March 1979. The 123s were built at Swindon Works in 1963 and were the final flowering of the BR first-generation DMU fleet. Towards the end of their careers, many were transferred to the Eastern Region and were formed into hybrid sets that included Class 124 'Transpennine' DMU cars.

Almost a total 1960s scene at Stafford on the evening of 19 September 1978, as Class 304 set 035 was awaiting departure with a Crewe to Rugby service. The rebuilt station at Stafford was a product of the massive investment and upgrade of the former LNWR Premier Line in the early to mid-1960s, when electrification was completed from Liverpool Lime Street, Manchester Piccadilly, Crewe, Birmingham New Street, and London Euston. The 304s (AM4 when built) shared the semi-fast services along the WCML south of Crewe with the AM10 (Class 310) EMUs. The 304s were built in two batches at Wolverton Works, totalling forty-five four-car sets.

In the Longdendale Valley, Class 76 *Tommy* 76046 was passing over the B6105 Glossop to Crowden road at Torside level crossing, while descending downgrade from Woodhead with loaded 21-ton coal hoppers and 16-ton coal wagons bound for the Manchester area on 15 August 1979. The *Tommy* had been one of a batch of thirteen named locomotives that were originally fitted with steam heating for working passenger trains between Manchester Piccadilly and Sheffield Victoria. As E26046, the loco had been named *Archimedes*, but following the withdrawal of passenger services over the Woodhead route in 1970, the name was removed. The loco had been built at Gorton Works in August 1952. Being vacuum braked only, 76046 was withdrawn from service in November 1980, some eight months before the 1,500-v DC overhead wires of the Woodhead route was closed in July 1981.

At London's lost terminus, a pair of Class 501 three-car EMUs were stabled between duties at a very rundown-looking Broad Street on 20 October 1979. Driving Motor Brake Second M75150 was the nearest car. The Class 501 fleet consisted of fifty-seven three-car units on short 57-foot underframes that were built at Eastleigh Works in 1955–1956 for use on the former LNWR DC electrified lines in North London (Euston–Watford and Broad Street–Richmond). The 501s were eventually replaced by Class 313s in 1985. The terminus at Broad Street was opened by the North London Railway in 1865. When most of the structure was cleared in November 1985, one single open-air platform was retained until June 1986, when all railway activity ceased here. The station site was prime real estate in the heart of the city of London.

In the lost world of the active port at Birkenhead Docks, Class 03 03162 was stabled for the weekend at Duke Street on 24 August 1984. The 30-ton diesel shunter had been constructed at Swindon Works in September 1960, numbered D2162. The loco was an Eastern Region-allocated machine for most of its working life but was eventually withdrawn from Birkenhead Mollington Street depot in May 1989. No. 03162 made it into preservation, but the sight and sound of a 03 pottering around the extensive dock system at Birkenhead has gone forever.

With the former Lancashire & Yorkshire Railway stone-built waiting room on the eastbound platform, Class 25/3 25301 was starting away from Bamber Bridge, having been held at signal PN486 adjacent to the level crossing, while hauling a short engineers consist to Blackburn on 8 August 1979. The loco was one of the final batches of Sulzer Type 2s that were built by Beyer-Peacock in Manchester, the loco being new to traffic in May 1966 numbered D7651. It was renumbered 25301 in January 1973. No. 25301 was withdrawn from service in December 1983 and scrapped at Swindon Works in September 1984.

Crew change at Derby with Class 45 45058, stood in platform 1 with the 07.25 Plymouth to Edinburgh service on Saturday, 5 August 1978. This was in the days when Derby was busy on Summer Saturdays with locomotive-hauled inter-regional holiday services. The 'Peak' had been built at Crewe Works as D97 in April 1961. It was renumbered 45058 in January 1975 and was withdrawn from service in September 1987. The loco was scrapped by MC Metals at Springburn in 1994.

With thirteenth-century Conwy Castle as a backdrop, Class 47/4 47500 *Great Western* was getting underway from Llandudno Junction with a London Euston–Holyhead service on 27 July 1979. The train was about to enter the unique tubular bridge that takes the North Wales main line across the tidal River Conwy. Designed by William Fairburn and Robert Stephenson and completed in 1849, the 463-foot (141-m) bridge is sandwiched between the castle and the Telford suspension bridge of 1826. Civil engineers' wagons and a brake van were stabled on the low-level siding, complete with a small LNWR hand-cranked jib crane at the quay.

Stabled for the weekend under the partial cover of the LNWR overall roof at Rugby on 17 May 1981 was Class 310 310059. The fifty stylish AM10 four-car EMUs were built at Derby between 1965–1967 and were the first EMUs to utilise the BR Mk 2 coach bodyshell. The fleet was synonymous with the southern end of the WCML and its massive 1960s upgrade. Eventually superseded by new Class 321s, the 310s were transferred to the London Tilbury and Southend line in the late 1980s. They were all withdrawn following thc introduction of new Class 357 EMUs between 2000 and 2002.

In a familiar scene that was once commonplace across the BR network, three Class 08 diesel shunters (08284, 08918, and 08922) were stabled over the weekend between duties in the busy yard at Warrington Arpley on 12 August 1984. Of the trio, 08918 and 08922 were both built at Horwich Works, emerging new as D4148 and D4152 in 1962, while 08284 was a product of Derby Works, new as 13354 in June 1957; it became D3354 in March 1962.

On a hot late Summer Saturday, Class 45 45075 was topping the 1 in 37-gradient of the Lickey Incline at Blackwell, almost down to walking pace, with the 08.25 Taunton to Newcastle service on 5 September 1981. The former Midland Railway Bristol to Birmingham main line was at the time very busy with loco-hauled inter-regional holiday trains.

A lucky well-timed patch of sunlight on an otherwise dull day was illuminating Southern Region Class 203 Hastings DMU 1032, as it was leading the Hertfordshire Railtours/ Southern Electric Group's 'The Long Thin Drag' excursion from London Victoria, passing through Hellifield, bound for Carlisle on 12 April 1986. Units 1032, 1011, and two power cars from unit 1001 running as a 6+6+2 coach formation were used on this unique and one-off tour that returned from Carlisle via the WCML.

A four-car Class 127 (Red Triangle) DMU, with Driving Motor Brake Second (DMBS) M51623 leading, was departing from Elstree on the Up Slow line with a Bedford to St Pancras all-stations service on 12 April 1980. There were originally thirty four-car Class 127 suburban sets built at Derby in 1959. The outer driving cars of each set were fitted with Rolls-Royce engines and had hydraulic transmissions. The sets were very intensively worked, and subsequently became very work-worn in their later years, soldiering on until electrification and the replacement by new Class 317 EMUs in 1983. Following their withdrawal, twenty-two driving cars were converted to eleven two-car parcel/mail units fitted with bodyside roller shutter doors in 1985.

Class 20 20094 (along with 20080 just out of sight) was propelling loaded HEA coal hoppers across the level crossing at Deepdale Mill Street, well off 'the beaten track' in the backstreets of Preston on 2 August 1985. This section of line, which had formed part of the original Preston & Longridge Railway route of 1840 to its passenger and goods terminus, was accessed from nearby Deepdale Junction; the short section of line had been retained to access the coal concentration depot at nearby Fletcher Road.

On former Great Western Railway territory, which was transferred to the London Midland Region in the 1960s, Class 47 47258 was arriving at the Herefordshire market town of Leominster with the 12.22 Cardiff to Crewe service on 29 July 1978. The loco had emerged new from the Brush Works at Loughborough numbered D1938 in April 1966 (Works No. 400). It was renumbered 47258 in December 1973 and clocked up over thirty-eight years in service, before being withdrawn in September 2004. No. 47258 was scrapped by Booth's of Rotherham in February 2005.

With the 1960s architectural 'splendour' as backdrop (corrugated iron sheeting and bare concrete), Class 86/2 86213 was stood at London Euston with a service for Wolverhampton on 21 October 1979. Period details visible are a Class 310 EMU in the left background, BRUTE trolleys, and another Class 86 with air-conditioned Mk 2 stock on the adjacent platform. It is now difficult to fathom why in the 1960s the former Premier Line ended up with a London terminus that was constructed with unpainted breeze blocks, bare concrete and corrugated metal sheets. The entrance hall/waiting area (with no seating) was at the time of its opening (in 1968) supposed to have the look, glamour, and sophistication of an airport terminal building, with the trains hidden from view. Even from our twenty-first century viewpoint, when some 1960s structures are seen to have some architectural merit, Euston is a stark reminder of what happens when there is no thought for the past or heritage. The perpetrators of this edifice destroyed the original London & Birmingham Railway terminus, designed by Philip Hardwick, complete with its great hall, its iron and glass overall roof, and the Doric Arch on Euston Road—all in the name of 'progress'.

Class 506 at twilight at Hadfield on 29 November 1984. The Class 506 1,500-v DC units were in their last few weeks of operation on the services from Manchester Piccadilly to Glossop and Hadfield. Entering service in June 1954, the eight three-car EMUs remained as the only 1,500-v DC trains in the UK after the rest of the electrified line to Sheffield was closed beyond Hadfield in July 1981. In 1984, the two rusty tracks beyond Hadfield, behind the 506, which in the 1950s formed part of the most modern main line in Britain, were still intact all the way through to Penistone.

With the WCML in the foreground, Class 25/1 25078 had arrived back at Wigan Springs Branch with an electrification maintenance train following Sunday engineering work on 1 April 1984. Above the twenty-five in the sidings at the rear of Springs Branch Depot were withdrawn locomotives 40158, 25141, and 25188. No. 25078 had been built at Darlington as D5228, which had been new in September 1963. It lasted twenty-two years in service before being withdrawn in September 1985. Berry's of Leicester converted 25078 into 73 tons of scrap metal in 1987.

Class 44 44002, formerly named *Helvellyn*, was stabled near to the entrance road at Toton Depot on 3 September 1978. The loco was one of the pioneer Class 44s (D1-D10) that carried the names of British mountains and subsequently became known as 'Peaks'. No. 44002 was built at Derby Works as D2 and had been new in September 1959. It initially was fitted with experimental gearing to run in excess of the 90-mph top speed and was tested with a short train on the WCML, reaching 110 mph. D2 went new to Camden Depot (1B), but the arrival of new Class 40s and the introduction of new Class 45s on the Midland Main Line saw the 44s transferred to Toton Depot in the early 1960s where they plied their daily trade on freight duties. The class were regular performers from the East Midlands to Whitemoor Yard in Cambridgeshire on coal and mineral workings. The loco was renumbered 44002 in March 1974, and although its official withdrawal date was 1 February 1979, it appeared to have already been stored when photographed on that sunny afternoon at Toton. No. 44002 subsequently went full circle; the 133-ton locomotive was scrapped at Derby Works in October 1979.

At the West Lancashire market town of Ormskirk, Merseyrail Class 507 507004 was awaiting departure with a service to Liverpool Central on 18 April 1979. The line from Liverpool to Ormskirk was electrified by the Lancashire and Yorkshire Railway with electric services commencing on 1 April 1913. Ormskirk was once a double track through station but was heavily rationalised in 1970, when the line to Preston was singled, with the services from Liverpool and Preston meeting 'head on' in the same platform against a pair of stop blocks located back-to-back. An emergency loop was kept *in situ* for many years to bypass the two terminal sections. There is much evidence in this view of redundant railway infrastructure at this once busy location, typifying the aftermath of 'rationalisation' as the network and associated facilities contracted.

With a BR 'Merrymaker' excursion from Stevenage to the Lancashire seaside resort of Morecambe formed of Mk I stock, Class 31s 31191 and 31233 were about to take the 'Little North Western' route to Wennington and Carnforth at Settle Junction on 13 May 1978.

A stored Manchester–Bury Class 504 EMU in an early version of BR Blue livery, formed of cars M77163 and M65442, was stood in what had been the entrance roads to the former Bury MPD (9M) on 14 August 1977. The electric car workshops are behind the unit. The 8.5-mile route between Manchester Victoria and Bury Bolton Street was electrified by the Lancashire & Yorkshire Railway in 1916, using a unique 'side collection' third rail system. Twenty-six two-car Class 504s were built at Wolverton in 1959 to replace the old L&Y sets. Service reductions on the line in the early 1970s saw several of the 504s placed into storage at Bury, including some sets in BR green livery.

Class 86/2 86232 was paused on one of the through roads at Crewe station during a crew change, while working a northbound mail/parcels service on 9 June 1979. A Class 304 EMU on a service to Liverpool Lime Street is visible on the left. The station wall with its iron canopy has been a backdrop here since LNWR days and has been silent witness to the comings and goings at this busy location. LNWR Jumbos and Claughtons, LMS Stanier Pacifics, BR Class 50 Diesels, and AC Electric locos are all now history, with the structure now playing host to Class 390 Pendolinos. No. 86232 had been built at Doncaster Works as E3113 in August 1965.

In what was a daily and frequent sight at the time, but is sadly now part of history, Class 56 56084 was plodding southbound along the Midland Main Line, approaching Chesterfield with a well-loaded coal working formed of HAA hopper wagons on 13 August 1985. The 56 had been built at Doncaster Works in October 1980 and lasted almost twenty-eight years in service, being withdrawn in September 2008. No. 56084 was scrapped by Booth's of Rotherham in March 2009.

A scruffy-looking Gloucester RC&W Class 128 motor parcels van, M55993, coupled with an ex-Southern Railway PMV van at the rear was stabled in the sidings at Shrewsbury on 23 June 1979. There had been ten of these non-passenger-carrying single units built in 1959–1960; six were gangwayed and four were non-gangwayed. A lack of work and surplus loco-hauled parcel/mail vehicles saw the last of the class withdrawn from service in 1991.

Above left: With six blue/grey Mk I coaches behind the loco, Class 31/4 31423 was stood on the Up Slow at Leyland station with the 15.47 Preston to Liverpool Lime Street service on 12 August 1989. This was one of several loco-hauled workings in the north-west provided due to a shortage of multiple units and the ongoing problems with the then almost new Class 142 Pacer fleet.

Above right: Several BRCW Class 104 DMUs were stabled at Buxton station on 20 August 1979. The nearest three-car set was formed of cars M50482, M59138, and M50460. Buxton depot had a sizeable allocation of these DMUs from the late 1950s until the early 1980s, mainly deployed on the busy and frequent commuter services to and from Manchester Piccadilly.

Tyseley-allocated Class 122 single-unit M55006 had arrived at Stourbridge Town via the steeply graded 0.8-mile single track branch from Stourbridge Junction on 18 April 1981. The railcar had been built by the Gloucester Railway Carriage and Wagon Company (GRCW) in 1958.

2

Eastern Region

With the overnight frost melting on the roof in the foreground, Class 13 'Master & Slave' shunters 13001 and 13003 were awaiting their next move in Tinsley Yard at Sheffield on 3 January 1978. There were three 'Master & Slave' locomotives created from six Class 08 shunters for hump shunting at Tinsley Yard in 1965. They were originally numbered D4500–D4502. No. 13001 was formed of D4190 and D4189, while 13003 was made up of D4188 and D3698. No. 13002 was the first to be withdrawn in 1981 as the activity in the yard began to decline. The closure of the hump yard in 1985 sealed the fate of the remaining two, 13001 and 13003, both of which were withdrawn in January 1985. No. 13001 was subsequently scrapped in May 1985 at Swindon Works, and 13003 was put to the torch at Doncaster Works in September 1986.

With the steam heat on, the driver of Class 37/0 37037 was looking back for the 'right away' at Ely with the 13.30 Kings Lynn to London Liverpool Street service on 2 May 1980.

Class 55 Deltic 55005 *The Prince of Wales's Own Regiment of Yorkshire* was arriving at York with a London Kings Cross to Newcastle service on 17 June 1978. The Deltic had been new from Vulcan Foundry at Newton-Le-Willows as D9005 in March 1961, and its first home was at Gateshead (52A). It was named in October 1963. The loco was renumbered 55005 in February 1974 and eventually clocked-up almost twenty years of ECML service; its official switch off date was 8 February 1981. No. 55005 was scrapped at Doncaster Works in February 1983.

A pair of worn-looking Clacton Class 309 EMUs 625 and 621 were stood on the Great Eastern main line at Colchester on 5 April 1981. The Class 309s were built at York in 1962–63 and were the first BR EMUs that were capable of 100 mph. The last of the twenty-three units were retired from Liverpool Street/Clacton-on-Sea/Walton-on-the-Naze services in 1994, but following a period in store, several sets were re-activated for services around Manchester, until they were eventually withdrawn in 2000.

Class 76 *Tommy* 76025 was the centre of attention during a photo stop at Penistone while working the LGCB 'the Easter Tommy' railtour on 21 April 1981. The *Tommy* had worked from Guide Bridge that had started out earlier in the day from Liverpool Lime Street behind Class 25s 25057 and 25083. No. 76025 would come off the train at Rotherwood Sidings near to Sheffield, where Class 31s 31116 and 31227 took the train forwards to Lincoln. The same locos were used on the return journey. No. 76025 and its exploits on 'the Easter Tommy' are now part of the Woodhead route history. It was the last ever Class 76-hauled passenger train and the last ever passenger train to pass through the 1954 'new' Woodhead Tunnel in both directions, before the route was finally closed in July 1981.

A pair of weary-looking Class 20s 20185 and 20190 were stood in the sidings at Thornaby Depot on 20 September 1992. No. 20185 had been built at Vulcan Foundry as D8185 in February 1967. It was renumbered 20185 in October 1973 and was officially withdrawn just a few days after this image was recorded, its actual switch off date being 2 October 1992. No. 20185 was scrapped by MC Metals at Springburn in August 1994. No. 20190 was also a product of Vulcan Foundry that had been new in January 1967 as D8190. It was renumbered 20190 in January 1973 and was eventually converted to a Class 20/3 and renumbered 20311 in 1998.

The contemporary BR Blue scene at Norwich on 27 April 1980. In the foreground was the station pilot, Class 03 03084 complete with its runner truck (ex-BR Conflat TDB530638) and coupled with a steam heat van warming a rake of coaches. Alongside was a BRCW Class 104 DMU, while in the station was Class 47 47184 *County of Cambridgeshire* on a service to Liverpool Street formed of early Mk 2 stock. Above the Mk 2 coaches were some parcels/mail vans including a BR CCT, while two rakes of Mk Is were in the sidings beyond. In the distance, there was a well-filled goods yard and sidings, with grain hoppers, 16-ton mineral wagons, oil tank wagons, and Prestflo cement wagons visible.

On a freezing winter afternoon, a Derby Class 114 DMU formed of cars E56022 and E50022 was running under clear signals at Worksop while working a Lincoln to Sheffield service on 16 January 1977. There were forty-nine two-car 114s built at Derby in 1956–57. They had 63.5-foot underframes and were often referred to as 'Derby Heavyweights'. The fleet went new to Lincoln depot and were synonymous with services across Lincolnshire and the East Midlands. The last units were retired from passenger use in 1992.

In the days when Selby was located on the East Coast Main Line, Class 40 40056 was taking the Down Through Line with a lengthy rake of Mk I stock on 12 September 1981. The train had just passed through Selby South Junction and was the 08.40 Manchester Victoria to Scarborough service. The 40 had been built at Vulcan Foundry, emerging new as D256 in January 1960. It was renumbered 40056 in March 1974. The loco was withdrawn with serious power unit damage in September 1984 and was subsequently scrapped at Doncaster Works in April 1985.

Class 47 47018, complete with its Stratford Depot trademark silver roof, was awaiting departure with the 12.30 Liverpool Street to Norwich service on 20 October 1979. In the background, a Class 308 is visible, along with various platform trolleys (including BRUTES) in Rail Blue. The 47 had been built at Crewe Works, as D1572 in April 1964. Renumbered 47018 in February 1974, the loco had over twenty-seven years in service, before being withdrawn in November 1991. No. 47018 was scrapped by Coopers Metals of Sheffield in March 1994.

The mortal remains of Class 08 diesel shunter 08276 (including one of its oval worksplates) was in the cutting area at the rear of Doncaster Works on 19 October 1979. The 08 had been built at Derby Works as 13346 in April 1957. It became D3346 in 1962 and was renumbered 08276 in 1974. It had been a Scottish Region-allocated machine for all of its twenty-year career, and was withdrawn from Dundee in June 1977. As the number of freight flows and yards declined, including vacuum-braked wagons, the BR Class 08 fleet was much reduced during the 1970s.

A Class 108 two-car DMU with Driving Trailer Second Lavatory (DTSL) E56193 leading, coupled to Driving Motor Brake Second (DMBS) E50607, was stood in platform 1 at Ilkley with a service to Leeds on 9 August 1980. The Wharfdale Line from Leeds to Ilkley was subsequently electrified in 1995.

With the driver walking up the platform, Class 302 set no. 268, was stood at the end of the London, Tilbury & Southend Line at Shoeburyness, ready to work the 39 miles to Fenchurch Street on 1 May 1980. There were 112 four-car 302 EMUs built at York and Doncaster between 1958 and 1960. The last sets were withdrawn from passenger service in 1999. There are some period motors just visible on the right of the 302, a couple of Hillman Hunters, a battered Commer Van, what looks like the front end of a Vauxhall Viva, and a Reliant Robin almost hidden behind it.

Class 08 08506 was stabled between duties alongside the East Coast Main Line at Darlington on 14 October 1984. The diesel depot that was opened in 1958 is visible in the background. The loco had been built at Doncaster as D3661 and was new in June 1958. It was renumbered 08506 in February 1974. It was finally withdrawn from service in February 2007 and scrapped by TJ Thomson of Stockton in July 2007.

Class 308 144 had reached journey's end at Braintree, having worked the 7 miles from the junction at Witham on 28 April 1980. Braintree has been the terminus of the line since the line beyond here to Bishops Stortford was closed to passenger traffic in 1952. The single-track branch was electrified in 1977. The 308 was one of the units transferred north in the 1990s to work the newly electrified Aire Valley services in West Yorkshire.

Class 55 Deltic 55021 *Argyll & Sutherland Highlander* was stood at York with an Edinburgh to London Kings Cross service on 19 June 1979. These were the closing years of the reign of these magnificent machines on long-distance East Coast Main Line Class 1 duties. The loco was new from Vulcan Foundry at Newton-Le-Willows in March 1962 as D9021 and survived in service until the last active day of Deltic ECML running on New Year's Eve 1981, completing nineteen years and nine months of running. No. 55021 was scrapped at Doncaster Works in September 1982.

Allocated to Sheffield Tinsley Depot, Class 37 37132 was stabled with a classmate, stood between duties in the winter sunlight at Worksop on a freezing cold 16 January 1977. The loco was built at Vulcan Foundry in 1966, numbered D6832; it was renumbered 37132 in 1973 and eventually became 37673. Withdrawn from service in 2000 after thirty-seven years in traffic, the loco was not scrapped until 2008. The now vanished sight of 16-ton mineral wagons well-filled with domestic coal (no doubt from a then active UK deep-level pit) are in the sidings on the opposite side of the line.

'Toffee Apple' Class 31/0 31103 was turning off the Great Eastern Main Line at Stratford in East London, while heading onto the North London Line, passing a container train bound for Felixstowe on 22 October 1979. The 31 appears to be hauling new Plasser & Theurer track machines that may have come from the Austrian manufacturer via the Harwich train ferry and were *en route* to the Plasser & Theurer UK premises at West Ealing for commissioning. No. 31003 was one of twenty early build Class 31s that were BR pilot scheme locomotives that had a red circle (electro-magnetic system) coupling code. The loco was built at the Brush Works at Loughborough, emerging new as D5503 in January 1958. It was renumbered 31003 in January 1973. The small number of locos in the batch with the unique coupling code made them non-standard; 31003 was withdrawn from service in February 1980 and scrapped at Doncaster Works in April 1980.

A Cravens Class 105 two-car DMU formed of cars E51255 and E56130 had arrived at the Suffolk market town of Sudbury, with a service from Colchester and Marks Tey on 5 April 1981. Much disused and abandoned infrastructure is visible in this scene, a legacy from the days when prior to 1967, Sudbury was a through station on a double-track route to Cambridge.

On a section of railway that has now disappeared entirely, Class 76 Tommy 76035 with its twin pantographs at full stretch was passing through the disused main line platforms at Penistone station with loaded 16-ton coal wagons bound for the Manchester area on 15 August 1979. The pair of platforms on the Woodhead route at Penistone were taken out of use when regular passenger services between Manchester Piccadilly and Sheffield Victoria were withdrawn on 5 January 1970. The Woodhead route between Penistone and Hadfield was closed in July 1981.

In refurbished livery, a Class 114 'Derby Heavyweight' two-car DMU, formed of cars E50001 and E56043, was awaiting to depart from a dilapidated looking New Holland Pier with a service to Cleethorpes on 28th September 1980. The linkspan connecting the station to the landing stage for the British Rail ferry service to Hull can be seen to the right of the station building. A Morris Minor and a Mark 3 Ford Cortina were parked behind the buffer stops. In this period only one track was *in situ* at this end of the pier. The platform on the left was used by road vehicles to and from the ferry. With the opening of the Humber Suspension Bridge, services along the 1,375-foot (419-m) long pier were withdrawn on 24 June 1981.

The goods shed at Halifax was providing refuge for Bradford Hammerton Street-allocated Class 03 03371 and its attendant runner truck on 9 August 1980. There was still sufficient work for a shunter at Halifax during the working week. This image was recorded on a Saturday afternoon when the loco had been 'put to bed' until needed on Monday.

A nocturnal scene at London Kings Cross with Class 55 Deltic 55009 *Alycidon* having arrived with a service from Newcastle on 18 November 1978. Alongside was Class 31 31193 having brought empty stock into the terminus.

Class 20s 20167 and 20009 were stabled between duties in the yard at York TMD on 31 May 1985. Also visible in this scene is 31207 and 08064. No. 20167 was among the last batches of 20s built, when production was restarted to cover for the unreliable and troublesome Clayton Type 1 (Class 17) fleet. The loco was built at Vulcan Foundry as D8167, new in October 1966. It was renumbered 20167 in January 1973. The loco was a Nottingham Division-allocated machine for all of its twenty-one-year BR career. Its withdrawal date was 9 November 1987, and it was scrapped by Berry's of Leicester in November 1988. Conversely, 20009 was among the first batch of English Electric Type 1s built, new from Vulcan Foundry as D8009 in October 1957. It went new to Devons Road depot in East London, and was a London area allocated machine until transferred north to the Nottingham Division in July 1969. It was renumbered 20009 in November 1973. No. 20009 was withdrawn following over thirty-one years of work, in July 1989. MC Metals at Springburn converted 20009 to 73 tons of scrap metal in November 1993.

A weary-looking Class 37 37172 was trundling through Ipswich with a lengthy train of ferry vans from Temple Mills Yard bound for Harwich Parkeston Quay (and the train ferry to Zeebrugge) on 28 April 1980. The loco had been built as D6872 and had been new in September 1963. It was renumbered 37172 in January 1973 and would become 37686 in 1986. It was withdrawn from service in July 2000, but not scrapped until 2006.

Class 306 EMU 043 was stood in platform 3 at Southend Victoria, as the lead unit of a six-car service to London Liverpool Street on 1 May 1980. Although normally associated with Liverpool Street–Shenfield services, the 306s also made visits to Southend. Originally an LNER design, ninety-two three-car sets were constructed by Metro-Cammell and BRCW and introduced into service in 1949. They were originally 1,500-v DC, but were converted to 25 kV in 1960–61. Well-worn bulk people carriers, the last of fleet was withdrawn in 1981. They were replaced by BREL York-built Class 315s, which have since been replaced by new Class 345 units.

A work-stained withdrawn Class 55 Deltic 55010 *The King's Own Scottish Borderer* was stood in the yard at Doncaster Works as part of the hastily arranged 'Deltic Farewell' open day on 27 February 1982. With its nameplates, worksplates, and cab front horns removed, the loco that was the first Deltic to clock-up 1 million miles in 1973 was awaiting its fate. The loco had been built at Vulcan Foundry as D9010 in July 1961. It was renumbered 55010 in June 1974 and the loco would last over twenty years in service, being withdrawn on Christmas Eve 1981, almost at the end of the Deltic era on the East Coast Main Line. Sadly, despite its milestone mileage achievement, 55010 was reduced to around 100 tons of scrap metal at Doncaster in May 1982. On that sunny winter day at Doncaster, fifteen surviving Deltics were on display—55002, 55004, 55005, 55007, 55008, 55009, 55010, 55011, 55013, 55015, 55016, 55017, 55019, 55021, and 55022.

Above left: At the former Cromer Beach terminus of the Midland & Great Northern Joint Railway (M&GNJR), a Cravens Class 105 two-car DMU, formed of E51294 and M56114, had arrived from Norwich on 29 April 1980. The driver was changing ends for the run to the small terminus at Sheringham. The station at Cromer had just one platform face in use, the station being surrounded by empty and redundant railway land—a legacy from the days when the station handled heavy summer holiday traffic from the north and the Midlands via the now closed M&GNJR system in North Norfolk.

Above right: In the days when Selby was on the East Coast Main Line and had Up and Down Through tracks, an eight-car Class 101/104 DMU, with Class 104 Driving Trailer Motor Composite (DMCL) E50566 at the rear, was forming a Leeds to Hull service on Saturday, 12 September 1981.

In the vanished world of industrial South Yorkshire, six Class 76 Tommies with pantographs raised were Sunday resting at Wath Depot on 5 May 1980. No. 76028 was the nearest locomotive. With rakes of 16-ton mineral wagons visible on the right, Wath-upon-Dearne had a colliery, and in the background, the smoke from the coking plant is visible; this area was a hive of industrial activity at the time and an important and very busy rail freight centre.

The view looking down into the yard at Leeds Holbeck Depot on 13 February 1987. There is a varied selection of motive power, including two 08s, Class 31 31250 and Class 45 45134. In storage were the Class 140 'Pacer' prototype, 140001, and doyen Class 141, 141001. The 141 was still in blue/grey (the blue being the 'Barrow Corporation' hue, as the Leyland National factory at Workington, where the bodies were assembled and finished, had no BR Rail Blue paint and selected the nearest shade they had in stock). Built in 1984, 141001 had never entered regular timetabled service and was the last of the class to remain in BR livery; it never carried the WYPTE green and cream colours.

In the days before the introduction of Class 56s on MGR coal workings, Knottingley depot had an allocation of hard-worked Class 47/3s fitted with slow speed control systems that were used on services to feed then-active strategic nearby coal-fired power stations at Drax, Eggborough, and Ferrybridge. No. 47376 was flanked by equally work stained sister locos stabled in the yard at Knottingley on 5 May 1980.

3

Southern Region

Having just run through the New Forest from Southampton on a fine summer evening, a pair of Class 423 4-VEPs—7748 and 7768—stood at New Milton with a London Waterloo to Bournemouth semi-fast service on 20 July 1986. The 4-VEP fleet consisted of 194 four-car units, mainly constructed at York Works. They were introduced in 1967, and production continued until 1974. They were a familiar sight on the Bournemouth line from 1967 onwards until the last examples faded away in 2005. No. 7748 was one of a batch of forty-nine units built in 1968–69, and at the time this image was recorded, it was allocated to Bournemouth Depot.

The one-time junction station for the Swanage branch at Wareham was reverberating to the tick-over of a scabby-looking Class 33/1 33106 while it was awaiting departure with the Weymouth portion of a service from London Waterloo on 24 July 1986. Behind the push-pull-fitted 33 was a pair of 4-TC sets that had been detached from a 3,300-hp 4-REP EMU at Bournemouth. This was a time-honoured sight along here and was soon to come to an end. Visible in the extreme left foreground between the tracks were new insulator pots ready for the extension of the third rail from Bournemouth to Poole and Weymouth that would eventually be completed in 1988. The loco was new from the Birmingham Railway Carriage & Wagon Company (BRCW) at Smethwick in August 1960 numbered D6519. It became a push-pull Class 33/1 in May 1967 with the Bournemouth line electrification and was one of nineteen of the class to be so modified.

In south-west London, 4-SUB 4650 was awaiting custom at North Sheen with a Waterloo to Waterloo 'Kingston Roundabout' working on 23 October 1979. The small station at North Sheen had been opened by the Southern Railway in 1930 and featured standard SR concrete components in its construction. In 1979, the station still retained its 1950s BR Southern Region green enamel signage. No. 4650 was one of the last batch of 4-SUB units that were built between 1947 and 1951 (units 4621–4754). The last of the well-worn 4-SUBs were finally retired in 1983.

4-EPB 5040 was departing from Charing Cross with a service to Caterham on 20 October 1979. The EPB units were built at Eastleigh from 1951 onwards to a Southern Railway design utilising SR construction jigs and SR type components (Sets 5001–5260). Later batches of EPB units had a different body profile based on BR Mk I Suburban Stock and featured flush cab ends.

A seven-car VEC-TIS Shanklin to Ryde Pier Head service with 3-TIS 033 trailing was departing from Brading on 25 July 1978. Complete with a red disc at the rear, the vintage units were leaving the end of the double track section from Sandown. These trains were sent to the 'modernised' Isle of Wight line in 1966–67. They were the first complete units to be turned out in the then new BR Rail blue livery. Forty-three ex-London Transport 'Standard' tube cars were refurbished for use on the 8.5-mile line. The four-car units were designated 4-VEC, and the three-car units 3-TIS. '*Vectis*' was the old Roman name for the Isle of Wight. Brading had been the junction for the short branch to Bembridge, which closed in September 1953.

Opposite above: Complete with a shunting pole across the buffers, Class 09 09024 was stood in the sidings at Poole with a vintage ex-Southern Railway 12-ton ventilated van on 24 July 1978. The van had been transferred to departmental service and was numbered ADB75254 M&Es RSE Section Woking—'To work between Woking & New Cross Gate'. The 09 had been built at the former Lancashire & Yorkshire Railway workshops at Horwich as D4112 in December 1961.

Opposite below: With the traffic keeping well out of the way, push-pull-fitted Class 33/1 33110 (complete with its orange flashing light attached to one of the front lamp irons and with 4-TC sets 426 and 429 in tow) was plodding along Commercial Road while negotiating the Weymouth Tramway with the 09.54 London Waterloo to Weymouth Quay Channel Islands boat train on 30 May 1986. The last BR-timetabled services to use the line along the streets from Weymouth station to Weymouth Quay was in 1987.

9024

91

Class 73/1 Electro-Diesel 73130 was arriving at Southampton Central with a short mail/parcels service on the evening of 28 July 1978. The ED had been built at Vulcan Foundry as E6037 and had been new in July 1966. It was renumbered 73130 in January 1973. The Southern Region had a total of forty-nine of these versatile locomotives that could be found on a variety of duties across the Southern network.

Class 09 09009 was stabled between duties at London Waterloo on 23 October 1979. The loco was one of the twenty-six 09s allocated to the Southern Region (09001–09026). It had been built at Darlington Works, numbered D3720, and had been new in April 1959. D3720 went new to Hither Green depot (73A), but during 1959, it also had spells at Faversham (73E) and at Norwood Junction (75C). The loco was renumbered 09009 in January 1973.

Class 423 4-VEP 7818 was stabled in the carriage sidings at Clapham Junction on 20 October 1979. In the right background was the Southern Region General Managers saloon DB975025, which was used on the rear of the royal wedding special from Waterloo to Romsey, hauled by Class 73 73142 in July 1981.

The pioneer Class 432 4-REP 2001 (ex-3001) was arriving at the New Forest town of Brockenhurst with the 09.00 Bournemouth to London Waterloo service on 24 July 1986. Behind the four-coach REP were two unpowered 4-TC sets 410 and 426. There were fifteen 4-REPs (3001–3015) built for the Bournemouth Line electrification in 1967, and they were used to haul unpowered 4-TC sets on express services between Waterloo and Bournemouth. With eight 400-hp electric motors per set and with a total of 3,200 hp on tap, the REPs were the world's most powerful third rail DC electric multiple units. When running on their own between depots or going for works attention, they were classed as a 'locomotive'. The last of the legendary REPs were withdrawn in 1991.

Class 33 33023 had just backed on to the ECS of the 12.10 Cardiff Central to Portsmouth Harbour service, at Portsmouth Harbour on 21 June 1986. No. 33056 had worked the train south from Bristol Temple Meads. No. 33023 had been built by the Birmingham Railway Carriage and Wagon Co (BRCW) at Smethwick in January 1961 as D6541. It became 33023 in February 1974. The loco was not withdrawn until May 2005 and was scrapped in November 2005.

Having arrived from Folkestone Harbour, Class 419 Motor Luggage Van (MLV) 68009 was stood in platform 2 at London Victoria coupled with a pair of 4-CEP units at 21.44 on 22 October 1979. There were ten of these battery-electric MLVs built at Eastleigh in 1959. They could operate off the live rail for around thirty minutes at low speed. The MLVs were withdrawn from service in 1991–92 with the end of boat trains to Dover and Folkestone. The entire class was transferred to departmental use and reclassified to Class 931s.

Complete with its paraffin tail lamp and guard with green flag, Class 405 4-SUB 4668 was awaiting departure from Richmond with London Waterloo to London Waterloo Kingston Roundabout service on 23 October 1979. No. 4668 was one of the last batch of 4-SUBs built between 1949 and 1951 (4621–4754). The last of the well-worn 4-SUBs were withdrawn in 1983.

Class 33/1 33108 was providing the power at the rear of a Weymouth to Bournemouth service formed of a non-powered 4-TC set 415, departing from Wareham on 24 July 1986. The 4-TC would be attached to the rear of a 4-REP EMU that would be waiting in the Up platform at Bournemouth ready for the onward journey to London Waterloo. In 1986, the extension of the live rails were underway from Bournemouth to Weymouth, and the uninstalled pots for the conductor rails can be seen in between the two tracks.

A four-car Class 508 508033 was departing from Dorking on 23 May 1982 with a service to Wimbledon. No. 508033 was transferred to Merseyside (minus a centre trailer coach) in 1984 and became 508133. The removed centre trailer coaches from the transferred forty-three Class 508s were inserted into new Southern Region Class 455/7 units (5701–5742 and 5750) to make them four-car units. In the centre of this scene is vintage Southern Region 4-SUB 4623, while on the left are sidings containing more Class 508s. The 'Odeon' type signal box was built by the Southern Railway in 1938.

The then sole remaining Hunslet Class 05, 05001, was fresh from a repaint and stabled with a rake of departmental wagons (including four ex-Southern Railway wooden dropside opens) in the engineers' sidings at Sandown on 25 July 1978. The 05 was shipped across the Solent to the Isle of Wight in June 1966 to work engineers' trains in connection with the electrification of the Ryde to Shanklin line and for maintenance purposes thereafter. Later allocated the departmental number 97803, which it never carried, 05001 was eventually retired in 1984 and replaced by Class 03 03079.

Class 416/2 2-EPB 5761 was stabled in number 3 road in the sidings at Orpington on 17 October 1982. The 416/2 units were based on the BR Mk 1 suburban bodyshell and were numbered 5700–5779. There were also fifteen similar EPB units built in 1954–55 for the South Tyneside Electric system. When the system was de-electrified in 1963, the units became part of the Southern Region fleet numbered 5781–5795. The last 416/2 EPBs were withdrawn in 1995.

With only another mile to go on its 172-mile journey, Class 33 33002 was arriving at Exeter Central with the 11.10 London Waterloo to Exeter St Davids service on 8 April 1980. The six-coach train was formed of four Mk 1 second-class coaches, a Mk I restaurant car, and a Mk 2 brake first corridor coach.

4-REP 3013 was providing 3,200 hp to the rear of a London Waterloo to Weymouth service departing from Southampton Central on 28 July 1978. The train was formed of two unpowered 4-TCs at the front, which would be detached at Bournemouth to be diesel-hauled to Weymouth. The 4-REP would be left behind at Bournemouth ready for the next working back to London. The scene here at Southampton is much changed—the signal gantry is long gone and the A3024 Western Esplanade dual carriageway now runs at the side of the line here.

On a grey winter day, Class 205 (3H) Thumper 1117 was crossing over to the Up Line as it departed from the terminus at East Grinstead with a service to London Victoria on 22 January 1984. The East Grinstead line was electrified in 1987, and the Bluebell Railway has had a platform at the other end of the station since 2013. Powered by a single English Electric 600-hp diesel engine mounted above the floor in the DMBSO Coach, 1117 was withdrawn in 1989 and subsequently scrapped.

4

Western Region

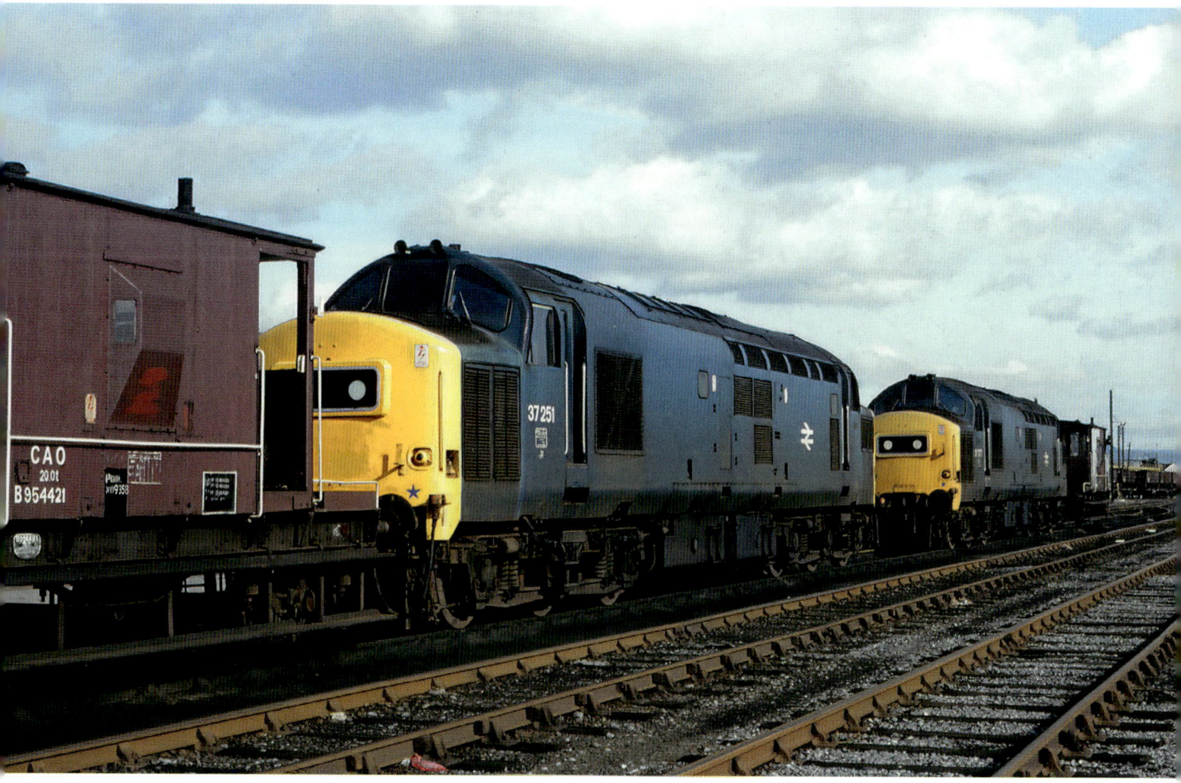

Class 37s 37251 and 37277 were whiling away a Sunday afternoon in the company of some BR 20-ton brake vans at Swansea East Dock on 11 October 1981. This was during the period when the vast and productive South Wales coalfields provided employment for locos such as these.

On a fine spring day, Class 47/4 47557 was getting underway from Dawlish with the 09.33 Penzance to Newcastle service on 29 May 1986. The 47 had been new from Crewe Works as D1591 in June 1964. It was renumbered 47024 in February 1974. The loco was selected for conversion to an ETH-fitted machine in 1979 and was renumbered 47557.

A heavily stripped Class 52 D1033 *Western Trooper* had reached the end of the line, huddled together with other doomed class members in the yard at Swindon Works on 22 February 1979. The loco had been built at Crewe Works, entering traffic in January 1964 in BR Maroon with small yellow warning panels. The loco was repainted blue in December 1967. It lasted a mere twelve years and eight months before being withdrawn in September 1976, clocking up an impressive 1,272,000 miles in its short working life. D1033 was finally reduced to 108 tons of scrap metal in April 1979. Judging by the painted inscription on the front skirt of the loco, some parts were to be removed for use on the preserved D1013. What fine locomotives these were, and what a waste that they were trashed with such haste.

In its original 'InterCity 125' livery, an HST with power car 43132 at the rear was descending the last few yards of Rattery Bank as the lead power car was entering the Up platform at Totnes with the 14.17 Plymouth to Newcastle service on 26 May 1986.

Formerly employed at the nearby Reading Signal Works, redundant Ruston and Hornsby 88DS (408493/1957) 97020 was awaiting disposal at the DMU depot at Reading on 15 November 1981. The loco had been replaced by ex-Scottish Region Barclay Class 06 06003 (97804). The 06 partially visible on the right of this scene was 06002, which was sent from the Scottish Region with 06003 as a source of spare parts. Unfortunately, the diminutive 17-ton Ruston was scrapped in August 1982.

A grimy and oil-stained Class 47 47246 was stood at St Austell with a lengthy Up parcels and mail working on 24 July 1980. The sidings to the right of the 47 were part of the then active Motorail terminal. The station footbridge in the background was adorned with an 'InterCity 125 They're Here!' banner.

In large logo blue livery, Class 50 50025 *Invincible* was taking the Down Through road at Dawlish Warren while working the 11.45 London Paddington to Penzance service on 29 May 1986. The Western Region lower quadrant signals were replaced by colour lights operated from the then new Exeter PSB in November 1986.

A Class 121 single unit W55026 (P126) and Class 25/1 25052 were stood in the yard at St Blazey on 27 July 1980. W55026 was one of sixteen 121s built by the Pressed Steel Company for the Western Region in 1960. Following departmental use and renumbered 977824, it passed into preservation but was scrapped in 2009. The 25 had been built at Derby Works as D5202 in May 1963. It was renumbered 25052 in January 1973. From being new, it had been a London Midland Region-allocated machine but was transferred to the Western Region at Plymouth Laira depot in May 1976. No. 25052 was withdrawn from service just under two months after this image was recorded, its official switch off date being 5 October 1980. The loco was consigned to history at Swindon Works in November 1980.

With the magnificent overall roof as a backdrop, Class 50 50047 *Swiftsure* was awaiting departure from London Paddington with the night sleeper service to Penzance on the evening of 19 October 1979.

The lost warship—in weather-faded BR Blue livery, Class 42 818 *Glory* was acting as gate guardian at the entrance to Swindon Works on 22 February 1979. The loco had been built at Swindon and had been new in March 1960. Following withdrawal from Plymouth Laira Depot in November 1972, having completed 1,137,000 miles in just over twelve years, the loco was unofficially preserved at Swindon. During its tenure there, 818 also carried BR maroon and BR green liveries. As the end drew near for Swindon Works in 1985, 818 was put to the torch in November 1985.

No. 31423 was stood at Bristol Temple Meads with a lengthy mixed ECS working on 25 October 1980. The 31 had been new from the Brush Works at Loughborough as D5621 in June 1960. It became 31197 in January 1973 and, following conversion to an ETH machine, was renumbered 31423 in 1975. It had just over thirty-six years in traffic before being withdrawn in September 1996. The loco was eventually reduced to 106 tons of scrap in September 2009. Class 31s were drafted in to the Western Region in the early to mid-1970s as replacements for the Class 35 Hymek diesel hydraulics.

Class 45 45061 was stood at Teignmouth with the 14.36 Paignton to Leeds service formed entirely of Mk I stock on 24 July 1979. Behind the front end of 45061 is the entrance to what was the Up Refuge line at Teignmouth; on Friday evenings in the busy 1950s, a rake of empty coaches would be berthed here overnight to form an early morning service on Saturday to clear large numbers of homeward bound holidaymakers. No. 45061 had been new from Crewe Works in May 1961 numbered D101. It became 45061 in July 1975, and following just over twenty years of service, it was withdrawn in July 1981 and scrapped at Swindon Works in April 1982.

Class 47 47085 *Mammoth* was arriving at Starcross with the 16.35 Paignton to Exeter St Davids service formed of six Mark 1 coaches on 26 May 1986. The 47 was one of a batch named by the Western Region in the 1960s with historical GWR names; 'Mammoth' was a name originally carried by a locomotive going back to the broad-gauge era of the GWR. The nameplates on these locomotives had lettering in GWR-style Egyptian Serif. No. 47085 was originally numbered D1670 and was a product of Crewe Works, entering traffic at Cardiff Canton (86A) in March 1965.

On 31 July 1980, Class 47089 *Amazon* was slowly backing a lengthy rake of Clayhood China Clay wagons into the sidings at Lostwithiel alongside the Cornish mainline. The station at Lostwithiel was once the junction for the branch to Fowey that was closed to passengers in 1965, but the branch has been kept active with China Clay traffic to the docks at Fowey.

With plenty of platform end observers watching, a work-worn Class 31 31124 was accelerating away from a signal check at Reading on the Up Through road while working empty newspaper bogie vans from Bristol, bound for Old Oak Common on 9 May 1979. Some 31s were allocated from the Eastern to the Western Region in the early 1970s as replacements for the Hymek Class 35 Diesel hydraulics.

Above left: A study in concentration in the now demolished 'A' Shop at Swindon Works on 14 May 1984—Class 08 08710 was having its yellow/black 'Wasp' stripes applied as its overhaul and repaint came to a conclusion. Not too sure what the health and safety people would make of this operation these days. No. 08710 was new from Crewe Works in April 1960 as D3877. It was a Scottish Region-allocated machine for most of its thirty-two-year operational life, being withdrawn in January 1993 and scrapped at Motherwell Depot in September the same year.

Above right: On the single-track Newquay branch, a three-car Class 118 DMU formed of cars W51305, W59472, and W51305 (P463) was arriving at the small station at Quintrell Downs with a Par to Newquay service on 28 July 1980. In charge here was a resident signal lady who manned the hand-operated level crossing gates across the A5038 road as well as the pair of lower quadrant home signals.

With the tide out, Class 45 45010 was heading west at Cockwood Harbour with the 07.28 Derby to Paignton service formed of Mk I stock on 23 May 1981. The loco had been built at Crewe Works as D112 and was new in August 1961. It was renumbered 45010 in December 1973. No. 45010 was withdrawn from service in March 1985 and was scrapped by MC Metals at Springburn in January 1989.

Above left: In its eye-catching and unique British Telecom yellow livery that had been applied in 1985, three-car Class 118 DMU (P460)—formed of cars 51317, 59469, and 51302—was setting out from the North Devon village of Umberleigh with the 17.14 Exeter Central to Barnstaple service on 25 July 1987. The Barnstaple line once formed part of the Southern Railway's 'withered arm' to North Cornwall. The route was transferred from the Southern to the Western Region in the mid-1960s.

Above right: In a scene that was once commonplace across the BR network, Class 08 08359 was shunting parcels/mail vans at Cardiff Central on 20 August 1983. The loco was built at Crewe Works and had been new as D3429 in March 1958. It went new to Bristol St Philips Marsh (82D) until January 1960 when it was transferred to Danygraig (87C). It appears to have spent most of its BR working life at various South Wales locations, becoming 08359 in March 1974. It was withdrawn from Cardiff Canton Depot just five months after this image was recorded, its official switch off date being 22 January 1984. The loco has survived into preservation at the Chasewater Railway.

On a gloomy winter day, Class 52 D1069 *Western Vanguard* was arriving at Bristol Temple Meads with the 08.05 Penzance to Liverpool Lime Street service (1M85) on 26 February 1974. D1069 had been built at Crewe and had been new in October 1963. It lasted almost twelve years in service, clocking up 1,168,000 miles before withdrawal in October 1975. D1069 was reduced to 108 tons of scrap metal at Swindon Works in February 1977.

Having crossed Brunel's Royal Albert Bridge from Cornwall, HST power car 43188 was entering the county of Devon with a SW/NE service on 31 May 1983. Sister power car 43187 was bringing up the rear. The background is dominated by the Tamar Suspension Bridge completed in 1961; it takes the busy A38 trunk road between the two counties of Devon and Cornwall. In the right foreground is the former GWR Royal Albert Bridge signal box.

Class 47/4 47446 was stabled with the stock of the Great Western Royal Mail travelling post office (TPO) service to London Paddington (1A01) at Penzance on the warm afternoon of 28 July 1979. Just like the TPO, 47446 is no longer with us. New from Crewe Works in April 1964 numbered D1563, the loco was in service for nearly twenty-eight years, being withdrawn from service in February 1992. No. 47446 was scrapped at Old Oak Common in 1997.

Flanked either side by London Transport Central Line tracks, Class 121 single-unit W55023 was stood in the single BR bay platform at Greenford with the 3.3-mile shuttle service to Ealing Broadway on the Great Western Main Line on 23 October 1979.

Among a plethora of Western Region lower-quadrant semaphore signals, Class 50 50046 *Ajax* was passing though Dawlish Warren with the 09.50 Edinburgh to Plymouth service on 10 April 1980.

A rather battered-looking Class 31/4— 31414—had backed on to a Saturday-only, seven-coach London Paddington to Barnstaple through service on 28 July 1979. This service was in the tradition of the 'withered arm'—Summer holiday trains from London working into deepest Devon on former Southern territory. Exeter St Davids was still controlled by GWR/Western Region lower-quadrant signals. Alongside 31414 is a loaded BRUTE trolley filled with parcels, destined for the vans stabled in the former Exe Valley bay platform.

The Down Motorail service from Kensington Olympia was stood at St Austell, having arrived behind Class 47 47157 on 27 July 1980. The Motorail car flats and their precious load have been detached from the rear of the Mk I coaches and are visible on the extreme left of this scene, as expectant holidaymakers walk along the platform to be reunited with their vehicles. Unfortunately, such civilised ways of taking your car on holiday to the West Country no longer exist.

With three admirers looking on, Class 45 45039 *The Manchester Regiment* had just been uncoupled from its rake of Eastern Region Mk 2 stock after arrival at Plymouth with the 07.35 Leeds to Penzance service on 1 August 1979. No. 45039 was withdrawn from service in December 1980 and scrapped at Swindon Works in May 1983. No. 45039 was replicated on many model railway layouts at the time, as Mainline Railways produced an OO scale model of the locomotive in BR blue in the 1970s and 1980s.

Class 33/0 33058 was heading west through Dawlish station on 28 May 1987 with a short consist from Exeter Riverside Yard to Tavistock Junction made up of a departmental track laying crane and an English China Clays (ECC) PBA bogie hopper.

5

Scottish Region

With the clock on the North British Hotel showing 12.55 p.m., Class 27/0 27033 was accelerating away from Edinburgh Waverley with the 12.51 service to Perth formed of Mk I stock on 23 July 1981. Getting in on the act with a simultaneous departure was a Derby-built Class 107 DMU with Greater Glasgow/Trans Clyde PTE branding, on the 12.52 service to Glasgow Central.

Class 311 311105 was departing from Glasgow Central with the 11.25 to Neilston via Queens Park on 23 July 1984. There were nineteen three-car Class 311s built by Cravens of Sheffield in 1967; they spent their entire lives on the electrified Glasgow suburban network until they were withdrawn in the 1990s. Two cars of set 311103 survive at the Summerlee Heritage Park at Coatbridge.

Class 06 06005 was stabled inside the depot at Dundee on 31 March 1979. This was in the days when the depot had a visitor's book and gaining access here was not a problem. Built at the Andrew Barclay & Sons workshops at Kilmarnock, the loco was one of thirty-five 204-hp locos built for the Scottish Region between 1958 and 1960 (D2410–D2444), with the remaining survivors eventually becoming 06001–06010.

With the train crew having 'opened' the level crossing gates on the approach to Leith Docks, Class 26 26007 was getting underway with a loaded coal working on 24 July 1981. The Scottish coal was destined for export via Leith Docks. The health and safety people would have a seizure if this happened now, but this was all in day's work in 1981. No. 26007 was the pioneer Class 26 Type 2 D5300, built by BRCW at Smethwick in 1958.

On an overcast summer day, Class 55 Deltic 55016 *Gordon Highlander* was raising the echoes as it passed through Markinch with the 05.50 London Kings Cross to Aberdeen service on 24 July 1981. The station was still furnished with BR 1950s Scottish Region pale blue enamel signage.

With the time at 11.12 a.m. on 23 July 1984, Class 20s 20149 and 20121 were preparing to depart from Glasgow Central with an excursion to Ayr. When photographed, both these Type 1s were allocated to Glasgow Eastfield Depot. No. 20149 had been built at Vulcan Foundry in 1966, while 20121 was built at Robert Stephensons and Hawthorns at Darlington in 1962. In the background is a rake of Sealink-liveried Mark I stock, used on services to Stranraer Harbour.

Class 27/1 27112 was awaiting departure from Dundee with a service to Edinburgh Waverley on 31 March 1979.

In blue/grey livery, Cravens Class 105 Driving Motor Brake Second (DMBS) SC51481 was passing through North Queensferry with a six-car mixed rake of DMU cars, bound for Edinburgh on 23 July 1981. The Cravens is formed into Haymarket set 105376, the other vehicles being Metro-Cammell Class 101 cars. Presumably, the Cravens had been finished in blue/grey to match the rest of the three-car set. SC51481 spent its entire working life in Scotland; it was one of a batch of twenty-two power/trailer sets supplied new in 1959. SC51481 would eventually be withdrawn from service in November 1981 and was scrapped by Mayer Newman at Snailwell, Cambridgeshire, in 1982.

Class 26/1 26036 was stood at Glasgow Queen Street awaiting departure with a service to Dundee on 14 June 1980. The loco had been built by the Birmingham Railway Carriage & Wagon Company (BRCW) at Smethwick in August 1959, numbered D5336. It was renumbered 26036 in January 1973 and clocked up just over thirty-four years in service before being withdrawn in October 1993. No. 26036 was consigned to history by MC Metals at Springburn in February 1995.

Proving that recessions are nothing new, four Scottish Region Class 20s—20199, 20028, 20198, and 20171, all surplus to requirements—were being stored undercover in the NCL goods depot at Dundee on 20 July 1981. All would eventually return to service, but all the locos in this view have since been scrapped.

The lost world of rail freight at Leith Docks—with a mixed freight departing the yard complex, including ICI Anhydrous ammonia tank wagons, 12-ton ventilated vans, and a brake van bringing up the rear, Class 08 08714 was stood between duties on 24 July 1981. The 08 had been built at Crewe in 1960, numbered D3881. It became 08714 in April 1974. The loco survived long enough to eventually become a Rail Express Systems (RES) machine, and on privatisation passed to EWS.

Following overnight snowfall, various DMUs were stabled at Hamilton Depot on 23 March 1980. The nearest car is a Class 116 Driving Motor Brake Second (DMBS), SC50892. The suburban-type Class 116s were built at Derby between 1957 and 1961. Hamilton Depot was located in South Lanarkshire, some 12 miles south-east of Glasgow. The depot—which had a substantial fleet of Class 101, 107, and 116 DMUs—eventually closed in 1982.

Into the last leg of its 393-mile journey from London Kings Cross, and with only 10 miles to go, HST power car 43111 was hammering past Prestonpans *en route* to Edinburgh Waverley on 24 July 1981. At the time, Prestonpans station was still furnished with BR 1950s Scottish Region light blue enamel signage.

Class 20 20146 was stood at Glasgow Central during track relaying work on Sunday, 23 March 1980. The 20 had been built at Vulcan Foundry, being new as D8146 in June 1966. It became 20146 in January 1973 and was eventually withdrawn from service in December 1988. It was scrapped by MC Metals at Springburn in April 1989.

Above left: In a daily scene that is now history, Strathclyde PTE-liveried Class 303 EMU 303075 was departing from Glasgow Central with a service to Newton on 23 July 1984. Class 87 87008 *City of Liverpool* was awaiting the start of its 401-mile journey with the 11.10 departure to London Euston. The ninety-one Class 303 units were built locally by Pressed Steel at Linwood between 1959 and 1961, for use on the Glasgow suburban network, which was electrified in 1960–1961.

Above right: On a rainy summer day, Class 101 three-car DMU 101325 with Driving Motor Brake Second SC51448 at the rear, had arrived at the terminus at North Berwick with a service from Edinburgh Waverley on 22 July 1981. Although having been rationalised to a single platform, the station still retained its British Railways 1950s Scottish Region light blue enamel signage. The North Berwick branch was eventually electrified in 1991.

6

BR Blue Miscellany

A regular sight across the British Rail network in the 1970s were the Leyland FG crew buses. In the days before everyone turned up in a white van, these yellow vehicles transported permanent way staff and their equipment to various work sites. JAU 202N (which had been new in 1975) was the regular that between jobs was parked at Bamber Bridge station, where it was recorded for posterity on 5 November 1983. The Leyland FG was a result of 'badge engineering'. The vehicles were a BMC product that had been originally a Morris/Austin design from the 1950s. Production of these Leyland light commercials had been moved to Bathgate in the 1970s.

The breakdown crane at Wigan Springs Branch, complete with its attendant runner trucks, was stabled under the partial cover of the old steam depot, keeping company with a Class 47 on the evening of 27 September 1977. TDM1001 had been built by Cowans Sheldon of Carlisle in 1931 for the LMS. The 45-ton steam crane was formerly at Lostock Hall depot near Preston, before being transferred to Wigan in 1972 following the cessation of locomotive activity at Lostock Hall.

In a scene that often occurred across the BR network as the system contracted, billed as the 'Last BR Train' along the Bury to Rawtenstall line, curious locals came out in force to witness the passage of 'The Rossendale Farewell' at the East Lancashire village of Ramsbottom on 14 February 1981. Class 104 DMCL M50505 was leading the six-car DMU formation slowly crossing over Bridge Street. Happily, this location is now part of the heritage East Lancashire Railway that runs from Heywood and Bury to Rawtenstall.

The corridor end of BR Mark I Brake Second Corridor Coach (BSK) 35453 was bringing up the rear of the 17.14 Blackpool North to Manchester Victoria service, stood at Leyland on 5 July 1990. The 37-ton coach had been built at Wolverton to diagram 181, Lot No.30721 in 1963.

BR Mk I Second Corridor Coach (SK) W18716 was in the formation of the 17.14 Blackpool North to Manchester Victoria service at Leyland on 23 July 1990. The coach had been built at Derby to BR diagram 146, Lot No. 30685 in 1962, numbered 25716. It had been renumbered 18716 in the early 1980s. The view shows the corridor side, and the coach was mounted on Commonwealth bogies. Through the worn rail blue paintwork can be seen the horizontal steel patches along the bottom of bodywork, no doubt applied at its last overhaul when corroded portions of the body were removed. Leaking roof gutters and windows allowed rainwater to run down behind the steel panels, and the accumulated water aided corrosion at the lower edge of the body against the underframe.

BR NPV 4-wheel van (ex-CCT) M94173 was stood in the yard at Aylesbury on 8 January 1984. There were 822 of these vans built at Earlestown Works between 1959 and 1961. They were fitted with end doors to carry cars/vans, but mostly used across the BR network as parcel/mail vans; examples could be seen from Cornwall to Scotland. By the mid-1980s, their use was on the decline. M94173 was finished in rail blue, and like most of its sister vehicles, it was in an unkempt and scruffy condition. When new, the vans were mostly finished in BR-lined maroon, some carrying maroon livery under a coating of grime into the late 1970s.

The 1958 BR Type 15 London Midland Region signal box is at Hest Bank on the WCML at the side of Morecambe Bay on 20 April 1978. Despite huge route modernisation and electrification completed in 1974, the signal box was retained to control the manual traditional level crossing gates, complete with wooden-boarded roadway that gave access to the shore at this popular location overlooking Morecambe Bay. Eventually replaced with lifting barriers, the signal box remained in use until May 2013, when control of the level crossing passed to the power box at Preston via CCTV. Notice the LNWR cast-iron notice just visible on the extreme right.

BR Mk 2 Brake Second Corridor Coach E17052 was at Southport on 28 August 1993 in the formation of a Manchester Victoria to Southport service. The coach was one of the last loco-hauled passenger-carrying vehicles on the system to retain BR blue/grey livery. It had been built at Derby in 1966 as a Brake First Corridor (BFK) to diagram 162, Lot No. 30756 (vehicles 14028–14055), its original running number being 14052. The coach was one of the first production Mk 2s built. The BR Mk 2 coach fleet (versions A/B/C/D/E/F) were built entirely at Derby between 1964 and 1975; they were a development of both the Mk 1 and experimental XP64 coaches. The Mk 2 fleet was an all-steel semi-integral design without any separate underframe, to increase strength and aid corrosion resistance.

A selection of the various types of ticket styles in use during the BR Blue era with a variety from all the regions. These were single-use, throwaway items—how many must have been issued, and how many have survived?

A stylish-looking BR brochure issued for the 1966 summer timetable that detailed various European services from the Southern Region, including the 'Golden Arrow/*Flèche d'Or*'. The cover illustration shows a British Rail-liveried cross-Channel ferry alongside an SNCF 67000-series diesel locomotive. The cover image is reminiscent of contemporary artwork for the catalogue covers of the 1960s Playcraft/Jouef HO scale model railway system.

As time progressed, the introduction of new stock, stock cascades across the system, and changing patterns of traffic saw familiar vehicles that not too long previously had been in frontline service heading for disposal. In what looked like a war zone, various redundant items of railway and industrial hardware were awaiting the torch in the then-extensive rural Bird's Scrapyard at Long Marston in Warwickshire on 12 October 1981. Left to right are three cars of London Transport CO/CP surface stock, a BR B4 bogie, two BR Mk I Composite Corridor Coaches (CK) 15838 and 15844, and a burnt-out Shenfield Class 306 EMU car from set 056. The pair of Mk Is had both been built by Metro-Cammell to Lot 30221 in 1956.

Above: With the ongoing contraction of wagonload freight services and the subsequent reduction of the diesel shunter fleet, many well-maintained serviceable locomotives were made redundant and bought by industrial concerns for further use. Complete with a shunter's pole resting on the front steps, former BR Class 03 D2049 was stood with a rake of MGR HAA hoppers at the loading point within the open cast coal mine at West Hallam in Derbyshire on 16 August 1981. The loco had emerged new from Doncaster Works on Christmas Eve 1958. Its first home was at Blaydon (52C), and it remained a North-Eastern Region-allocated machine for all of its twelve-year BR life. The loco was withdrawn from Goole in 1971. Surviving in NCB service for fourteen years, D2049 was scrapped in November 1985.

Below: In a time-honoured scene, the lady crossing keeper was closing the gates that protected the A5038 at Quintrell Downs in Cornwall on 28 July 1980. The single-platform station is the last call for local trains on their way to Newquay, just under 2 miles away. In 1980, this location still had lineside telegraph poles, Western Region lower-quadrant signalling, and the crossing gates here still had oil lamps with red lenses mounted on the top for protection in darkness. While the Newquay branch had a regular local stopping service to the Cornish main line at Par, on summer Saturdays, there were a number of long-distance inter-regional holiday trains and a through service to and from London Paddington.

Above left: Despite the BR corporate image, many vintage items of rolling stock survived in use by being transferred to engineers/departmental use, ensuring that often venerable vehicles could still be seen in use across the network. A former Southern Railway design 'pill-box' 25-ton brake van, still carrying its BR bauxite livery from the days when it was in service as a standard freight vehicle, was formed into a consist of prefabricated track panels at Warrington Bank Quay on 30 June 1979. DS55148 had been built at Ashford in 1948 and was still active on the network until 1989.

Above right: During the late 1950s and into the 1960s, many stations were equipped with updated platform lighting. Usually, the fittings containing fluorescent tubes were encased in a cover that had the station name printed on the outside. A pleasing touch was that the 'name' was printed in the then BR regional colours, dependant on which region the station was located. With time, presumably as a cost-saving measure, and to give an overall 'corporate image' nationwide, the lettering was produced in black only. With modern station lighting upgrades, many of these lights have been replaced since the 1980s, but a few still survive, such as this one atop a concrete post at the seaside terminus at Cleethorpes in Lincolnshire in November 2016.

As new coaching stock was introduced across the system, some vehicles that were still in good condition were modified and transferred to departmental/engineering duties. This was 25-kV Overhead Line Maintenance Staff and Office Coach ADB975744 stood at Preston as part of a wiring train on 10 August 1993. The vehicle had been a former BR Mk I Second Corridor Coach (SK) numbered M25440, which had been built to Lot No. 30350 at Wolverton Works in 1957.

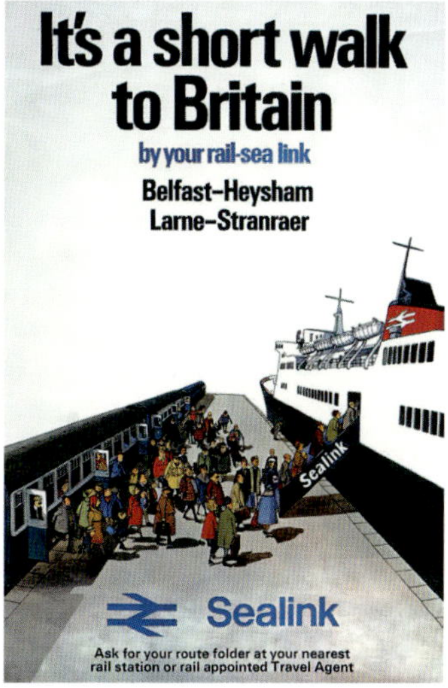

A BR/Sealink poster from 1970 promoting routes to Northern Ireland from Heysham and Stranraer. The illustration shows blue/grey Mk 2 stock as well as a BR Sealink ferry, complete with a reversed British Rail arrows logo on its funnel. For the BR shipping fleet, the normal arrows symbol made a ship look as though it was going in reverse, and a revised forwards facing design was implemented.

Well off the beaten track, the Holyhead Breakwater Railway was an isolated and unconnected 1.48-mile part of the British Rail network. The line ran from a quarry and along what was the longest harbour breakwater in the UK, for maintenance purposes. The railway was home to a pair of Andrew Barclay-built 153-hp diesel shunters dating from 1956. Originally numbered D2954 and D2955, they were renumbered 01001 and 01002 in 1973. Both locos carried their original BR 1950s black livery until they were taken out of service in 1979 (01001) and 1981 (01002). The railway was officially closed in 1980, when road vehicles took over maintenance duties. Unfortunately, both locos were scrapped on site in 1982.

At the former Lancashire & Yorkshire Railway workshops at Horwich, a small fleet of six redundant BR 16-ton mineral wagons were finished in rail blue for internal use at the then still busy works. The nearest wagon was formerly B576498 and was loaded with scrap metal on 9 May 1982.

In worn/faded blue and grey livery, Mk I Full Brake NEX 92369 was stood at Doncaster station on 27 April 1996. The 31-ton coach had been built by Pressed Steel in 1957 to diagram 711 Lot No. 30162, originally numbered 80960. It eventually became a dual-braked/ETH fitted coach with BR Commonwealth bogies, capable of 100-mph running. It was renumbered 92369 in the 1980s. With the reduction of loco-hauled trains and the reduction of parcels/mail traffic, a huge number of full brake coaches became redundant. No. 92369 saw out its last days as a passenger transfer vehicle at Doncaster station; should the station lifts be unavailable, it was used for the movement of any passengers with mobility issues between various parts of the station.

7

Resurgence and Epilogue

The BR blue era is now recreated in preservation and with heritage traction/rolling stock that is registered for use on today's national network. It is a period that is now fondly remembered and very much missed, from when we were all a lot younger.

Bibliography

Bridge, M., *Atlas of Mainland Britain*, second edition (Sheffield: Platform 5 Publishing 2012)

Haresnape, B., *British Rail 1948–78: A Journey by Design* (Shepperton: Ian Allan Ltd 1979)

Baker, S, K., *Rail Atlas of Britain*, second edition (Oxford: Oxford Publishing Co. 1978)

Butcher, Fox, Hall, R. P. P., *Departmental Coaching Stock* fourth edition (Sheffield: Platform 5 Publishing, 1990)

Website: BR Database—Complete Locomotive Database 1948–1997